STARK LIBRARY

STRONG KIDS HEALTHY PLATE
Great Grains

Katie Marsico

Published in the United States of America
by Cherry Lake Publishing Group
Ann Arbor, Michigan
www.cherrylakepublishing.com

Content Adviser: Debbie Fetter, PhD, Assistant Professor of Teaching Nutrition, University of California, Davis
Reading Adviser: Marla Conn, MS, Ed, Literacy specialist, Read-Ability, Inc.

Photo Credits: ©Casezy idea/Shutterstock.com, front cover; ©S_Photo/Shutterstock.com, 1; ©fizkes/Shutterstock.com, 4; ©symbiot/Shutterstock.com, 6; ©KK Tan/Shutterstock.com, 8; ©USDA/ChooseMyPlate.gov, 10; ©hanif66/Shutterstock.com, 12; ©SGM/Shutterstock.com, 14; ©Markuso/Shutterstock.com, 16; ©Brian Mueller/Shutterstock.com, 18; ©Wallenrock/Shutterstock.com, 20

Copyright @2021 by Cherry Lake Publishing Group
All rights reserved. No part of this book may be reproduced or utilized in any form or by any means without written permission from the publisher.

Cherry Lake Press is an imprint of Cherry Lake Publishing Group.

Library of Congress Cataloging-in-Publication Data

Names: Marsico, Katie, 1980- author.
Title: Great grains / Katie Marsico.
Description: Ann Arbor, Michigan : Cherry Lake Publishing, 2020. | Series: 21st Century Basic Skills Library: Level 3. Strong kids healthy plates | Includes bibliographical references and index. | Audience: Grades K-1. | Summary: "Grains give you energy to think and play. Learn about the different types of grains farmers grow and how they help you maintain a healthy diet. Content encourages balance and making healthy choices. This level 3 guided reader is based on the U.S. government's diet recommendations. Readers will develop word recognition and reading skills while learning about food and where it comes from. Includes table of contents, glossary, index, author biographies, and word list for home and school connection"—Provided by publisher.
Identifiers: LCCN 2020002748 (print) | LCCN 2020002749 (ebook) | ISBN 9781534168664 (hardcover) | ISBN 9781534170346 (paperback) | ISBN 9781534172180 (pdf) | ISBN 9781534174023 (ebook)
Subjects: LCSH: Grain in human nutrition—Juvenile literature. | Grain—Juvenile literature. | Nutrition—Juvenile literature. | Health—Juvenile literature.
Classification: LCC QP144.G73 M37 2020 (print) | LCC QP144.G73 (ebook) | DDC 613.2/731—dc23
LC record available at https://lccn.loc.gov/2020002748
LC ebook record available at https://lccn.loc.gov/2020002749

Cherry Lake Publishing Group would like to acknowledge the work of the Partnership for 21st Century Learning, a Network of Battelle for Kids. Please visit http://www.battelleforkids.org/networks/p21 for more information.

Printed in the United States of America
Corporate Graphics

Table of Contents

5	What Are Grains?
11	An Important Part of Your Plate
13	How Do Grains Help?
15	Dig into a Healthy Diet!
22	Glossary
23	Home and School Connection
24	Find Out More
24	Index

About the Author

Katie Marsico is an author of nonfiction books for children and young adults. She lives outside of Chicago, Illinois, with her husband and children.

Have you ever seen red popcorn seeds?

What Are Grains?

Do you eat oatmeal? How about popcorn or pasta?

These foods are made from **grains**. A grain is a small, hard seed.

Farmers grow many types of grains. Wheat, oats, and rice are some. Rye and barley are others.

Cereals and breads are made from grains. So are pasta and rice. Crackers and cookies are also grain foods.

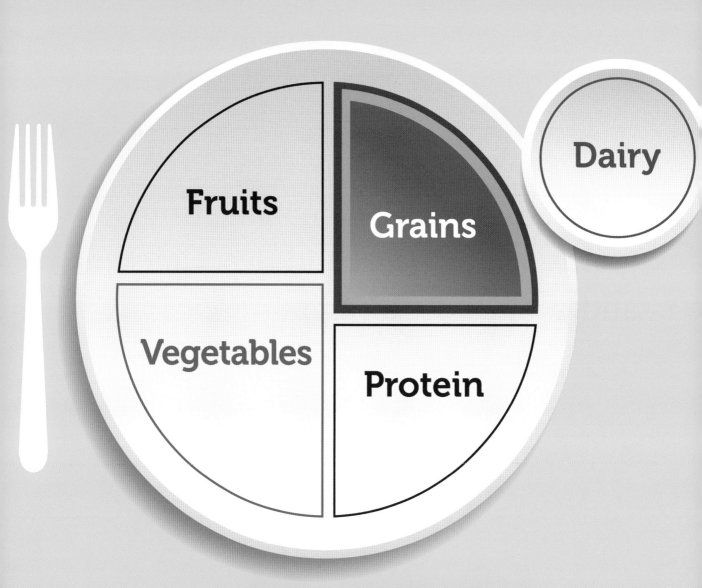

An Important Part of Your Plate

Grains are one of five main **food groups**. These groups make up a **balanced diet**.

You need a balanced diet to be healthy. Good food choices help your body grow.

Count how many food groups are on the plate.

How Do Grains Help?

Grains add **fiber** to your diet. Fiber helps keep your heart healthy.

Grains also give you **B vitamins**. B vitamins give you **energy** to think and play.

Dig into a Healthy Diet!

You should eat 5 to 6 ounces (.15 to .18 liter) of grains each day. A slice of bread is about 1 ounce (30 milliliters). So is 1 cup (.24 liter) of cereal.

Half of your grain foods should be **whole grains**. Whole grains contain fiber, vitamins, and minerals.

Oatmeal and brown rice are whole-grain foods. So are **cornmeal** and **bulgur**.

Refined grains don't contain the whole grain **kernel**. This helps them last longer than whole grains. But it also makes refined grains less healthy.

White flour and white bread are refined grains.

Do you help plan meals in your home? Share what you've learned about grains. What tasty, healthy choices will you make? What new grain foods will you try?

Glossary

B vitamins (BEE VYE-tuh-minz) substances found in grains that help turn food into energy

balanced diet (BAL-uhnsd DYE-it) eating just the right amounts of different foods

bulgur (BUHL-gur) a type of dried, cracked wheat often used in Middle Eastern cooking

cornmeal (KORN meel) dried corn that is mashed into tiny pieces

energy (EH-nur-jee) the ability or strength to do work

fiber (FYE-bur) a substance found in grains that helps the body break down food

food groups (FOOD GROOPS) groups of different foods that people should have in their diet

grains (GRAYNZ) the seeds of certain grasses that farmers grow for food

kernel (KUR-nuhl) the inner, softer part of a seed

refined grains (rih-FINED GRAYNZE) grains or grain foods that do not contain the whole grain kernel

whole grains (HOLE GRAYNZE) grains or grain foods that contain the whole grain kernel

Home and School Connection

a	day	heart	oatmeal	small
about	diet	help	oats	so
add	dig	home	of	some
also	do	how	on	strong
an	does	important	one	tasty
and	don't	in	or	than
are	each	into	others	the
balanced	eat	is	ounce	them
barley	energy	it	part	these
be	ever	keep	pasta	think
body	farmers	kernel	plan	this
bones	fiber	last	plate	to
bread	five	learned	play	try
brown	flour	less	popcorn	types
build	food	longer	red	up
bulgur	from	made	refined	vitamins
but	give	main	refresh	what
cereal	good	make	rice	wheat
choices	grain	many	rye	white
contain	groups	meals	seed	whole
cookies	grow	milk	seen	will
cornmeal	half	minerals	share	you
count	hard	my	should	your
crackers	have	need	six	you've
cup	healthy	new	slice	

Find Out More

Book

Black, Vanessa. *Grains*. Minneapolis, MN: Jump! Inc., 2017.

Website
USDA—MyPlate Kids' Place
www.choosemyplate.gov/browse-by-audience/view-all-audiences/children/kids
Use games and learning activities to find out more about grains and healthy eating.

Index

barley, 7
bread, 9, 15, 19
bulgur, 17

cereals, 9, 15
cookies, 9
cornmeal, 17
crackers, 9

diet, balanced, 11, 15–21

energy, 13

fiber, 13, 17
flour, white, 19
food groups, 11

grains
  how much you should eat, 15
  how they help, 13
  refined, 19
  types of, 7
  what they are, 5–9
  whole, 17

minerals, 17

oatmeal, 5, 17
oats, 7

pasta, 5, 9
popcorn, 4, 5

rice, 7, 9, 17
rye, 7

vitamins, 13, 17

wheat, 7